Fire is Hot

Fire is Hot

Joanne H. Stroud, Ph.D.

Illustrations by Tashasan

Fire is hot.

Just try standing too close to a raging bonfire,

and you will know its strength.

One of the first lessons we learn in life is

not to play with fire.

Fire, of all the elements, is the one that most changes whatever crosses its pathway or comes in contact with it. Fire melts a solid bar of chocolate into a runny, gooey mass, for instance.

Fire changes

water

into steam.

This process makes many things possible,

from making a simple cup of tea

to making a sturdy steam engine.

Fire, steady and hot,

allows the making of pots and utensils.

When put to the fire in a kiln,

a vessel of soft clay emerges

as a solid object

able to hold water

or oil.

Fire makes possible

the baking of bread

from various grains. Dough

becomes edible when fire is applied.

Fire fuses metals together.

Mineral ores are brought together in a homogenous mass by smelting and forging.

Fire also is needed to transform

sand into glass. Think of how it would

be to do without all the glasses

and

bottles we use,

if we didn't have fire.

*F*ire destroys impurities.

Fire sanitizes much of what we eat,

killing the germs and bacteria.

All of the vegetables and meat we eat at meals

are made appealing by boiling, baking,

broiling, grilling, and

using fire in other ways.

Fire warms us

when we are cold.

Having a campfire to sit around makes a cozy time of an outdoor experience. Just staring at a fire, with its constantly changing nature, usually brings forth tales of adventures. Our imaginations are stirred as the coals of a fire are stirred. We say that we are "fired up" when creative juices are flowing.

Fire can be started

by rubbing two sticks together,

but this is hard to do as anyone who has

ever been in the woods and tried to do it will

tell you. Fires also can be started when

flint stones are rubbed together briskly

in order to get a spark.

Fire
is many colors.

Although we think of it as mostly red,

it is also blue when especially hot, or yellow

at the tip of a candle. Sparks leap like yellow gnats

when a fire crackles or hot metal is struck.

Rubbing amber also produces sparks.

Fire turns iron into steel.

In fact, this process needs all four

of the elements

—earth, air, fire, and water—

to make it happen.

"In the beginning, the iron is of the earth, still unborn, still unformed. Air rushes into fire. Fire renders hard, immutable earth into transitory softness. Water swallows earth as the iron is quenched, and the cycle is complete,"

                         is how Will R. Cummings describes it.

Iron is an important part of our planet: It is the fourth most abundant substance in the earth's crust and eighty percent of the earth's core underneath the surface. Iron also runs through our bodies in the hemoglobin of the blood stream which carries the necessary oxygen to all parts of the body.

Fire

in the body

is called fever.

The body uses fever's elevated temperature

to fight invading bacteria.

Our bodies combat disease

by withholding the iron needed

by bacteria to multiply.

Fire as heat is needed

for all things to grow.

The sun's fiery rays reach all the way to earth.

When a dense cloud cover obscures the sun's fire, all existing life suffers.

Fire can be explosive.

When joined with gunpowder

or with nitroglycerin in dynamite,

fire ignites the explosive mixture.

Fire seems related to electricity,

although it isn't directly.

Electrical charges run through the wires

of our houses.

When wires become overloaded, they produce heat.

They can even produce fires.

Fire in the heart is love,

or electrical-like attraction.

The Austrian poet Rainer Maria Rilke
reverses the usual order in using fire imagery:

"To be loved

means to be consumed in the flame;

to love is to shine

with an inexhaustible light."

W. B. Yeats, the great Irish poet,
often uses images of fire
    in his poems.
When questioning whether it is love
    or merely attraction
        in "The Mask,"
    he asks:

"What matter, so there is but fire
        In you, in me."

In another Yeats poem,
> with the use of fire imagery to speak
> of its transforming effect,
the speaker says of the two lovers:

> *"Yet the world ends when these two things,*
> *Though several, are a single light,*
> > *When oil and wick are burned as one."*
>
> ("Solomon and the Witch")

Fire is often part of religious festivals.

Most churches still make use of fire by lighting candles on the altar. Worshipers in Catholic churches light small candles to help their prayers. Candlelight on the dining room table is always considered romantic. It's hard to realize that once all reading and writing at night were done by candlelight. A lighted candle is often compared to the vulnerability of a person's life, because of the ease in snuffing it out. Macbeth in Shakespeare's play laments:

"Out, out brief candle!
Life's but a walking shadow."

It flames up from the sacred bush in the story of Moses. It is the wall of fire that is the final cleansing of sin at the top of Mount Purgatory in Dante's *Divine Comedy*. Traditionally, it is the flames of Hell that punish those who are never able to repent. But in Dante's *Inferno*, ultimate sinners are trapped in ice.

Fire burns upward,

suggesting ascent.

Our thoughts naturally drift up,

trailing from the base blackness

of the wick or log,

through smoky confusion,

finally rising and breaking

through to clear awareness above.

Fire makes us think

of both flame and ash.

Fire lends itself to myth

because it has the properties of the most extreme of opposites. Just think of its gifts — cooking and warming us on a freezing day — but also its dangers: able swiftly to demolish a dry pile of leaves that bursts into flame or a dangling sleeve that hangs over a fire.

"Myths are made up of actions that contain the opposites of themselves,"

Roberto Calasso reminds us.

That startling combination is what makes myths

so intriguing through the ages.

*F*ire as a halo surrounds the heads of many saints and martyrs, as they are depicted in paintings and statues.

In like manner, even in pagan times, Homer's tale of the Trojan War, **The Iliad**, describes the Greek hero Achilles as being surrounded by fire when he is fighting fiercely.

*F*ire represented the divine in many of the earliest mythologies and religions.

The most powerful gods were often sun gods. They were usually male gods, while feminine goddesses were related to the moon. Ra, the Egyptian supreme being, was a sun god. Whoever was king at the time received power from Ra and was actually an earthly manifestation of the sun god. Ra was often depicted with long arms representing the sun's rays stretching toward earth with little hands on the ends of the rays reaching downward to touch humans. For the ancient Greeks, Helios drove the sun in a chariot across the sky from east to west each day. Heliocentric means having the sun as a center.

Fire was stolen from the gods of Olympus by the Titan Prometheus in Greek mythology.

I am sure you will remember the story of Prometheus's theft of fire from the gods who owned it before humans did. Either because he was mad at Zeus and wanted to defy the Olympians, or because he felt sorry for humans and wanted to help them, Prometheus took some of the divine fire in a slender reed and carried it to earth. Humankind was, of course, quite grateful, but Zeus was furious, as having fire was one of the main ways that the gods were superior to the human race. The punishment that Zeus inflicted on Prometheus was severe. Prometheus was chained to a remote rock, where every night an eagle ate part of his liver. He was finally freed by Hera's son, Hercules, in one of his Twelve Labors.

Fire in Sanskrit is agnih.

In the Vedic tradition in India,

Agni is god of fire and guardian of man.

From this source and the Latin ignis,

we get our word "ignite."

*F*ire breaks forth from a log in Gerald Manley Hopkins' poem "The Windhover." Hopkins uses strange words and rhythms that sound vaguely like old English to describe the brilliant joy that floods the heart when suddenly we become aware of God's presence in all of nature. Out of a burned-out, sooty, black log, suddenly the glowing, fiery inside, hidden before, bursts into the open.

What glory is revealed that was previously concealed! Hopkins say it thus:

> "blue-black embers, ah my dear,
> Fall, gall themselves and gash gold-vermillion."

It is much the same in another Hopkins poem, "God's Grandeur."

Here we have another fire image that reveals God to us:

> "The world is charged with the grandeur of God.
> It will flame out, like shining from shook foil."

*F*ire is still used to celebrate a holiday

or joyful occasion.

Fireworks bursting make dazzling patterns in the night sky

and lift spirits high. The Chinese New Year or the Fourth of July

would hardly be as festive without fireworks. Candles make

jack-o'-lanterns shine eerily in the dark on All Hallow's Eve, the

earlier name given to Halloween, or on the night before

El dia de los Muertos, the Day of the Dead, in Mexico, for

instance. Cemeteries there are aglow with candlelight all

through the night, which is November 1st instead of October 31st.

Fire can be very destructive.

Under drought conditions, especially in the summertime, forest fires spread rapidly and are hard to contain or put out. They can burn for weeks. Millions of acres of forest are destroyed each year by fires that are started by carelessness in the use of matches or by lightning striking dry wood.

Fire as molten lava can engulf whole communities.

On the Bay of Naples, two of the most famous cities of the ancient world, Pompeii and Herculaneum, were covered with lava by the volcanic eruption of Mount Etna in 79 A.D. On the islands of Hawaii, Pele is the goddess of fire and volcanoes who spews forth her energetic fire from the molten core of the earth. Pele can slumber for many years until the earth stirs her to new life. Mount St. Helens in Washington has recently showed clouds of steam. New islands come into existence by lava from volcanoes in the seas. Atlantis was an island, possibly real, possibly mythical, that was destroyed by eruption of the earth's surface.

Fire has produced many mythical creatures.

Salamanders were part of the alchemical process in the medieval ages. This little lizard-like animal was able to live in a fire unharmed. So, like gold, it had a never-ending life. The phoenix with its bright red and gold plumage was thought to live a thousand years. When it was the time to die, the phoenix prepared its own funeral pyre, and the sun's rays set fire to it. A new phoenix always arose from the ashes, making it seem to be a creature of eternal life. Fire dragons are probably the most famous of these storied animals. Many of these dragons were guardians of treasure, such as in the story of Beowulf. Fire dragons are part of the Western tradition. In China, dragons were more friendly. They were either water creatures or fertility ones.

Fire **is the chief element**

necessary to bring about the final results

in alchemical experiments in the Middle Ages.

Actually all of the elements come into play. Common earth is the first material, or prima materia. Water is used for dilution and air for drying, but fire under the glass beaker is needed at several steps along the way. Drawings from this period often show a salamander, the imaginary fire creature just mentioned, inside the beaker making it all happen. A refined gold with eternal life was the goal.

Gaston Bachelard,

a French philosopher, wrote

The Psychoanalysis of Fire

in 1938, changing in so many ways

how we currently think of fire.

He shocked his readers by applying what was then the new psychological method of talking about needy, suffering people and instead analyzed an element of the natural world. These are the kinds of things he said:

> "That fire should be the sign of sin and evil is easy to understand…"

He was referring to descriptions of Hell and pictures that represent the devil with his tongue of fire.

But if fire represents evil in so many ways,

how did fire manage also to become

a symbol of purity?

"Fire purifies everything. . . .

Fire is all-purifying

because

it suppresses

nauseous odors."

Bachelard explains that it was not only that "cooked foods gave more strength to the men of a tribe who, having won the secret of fire for cooking, were better able to digest the prepared food, and thereby made stronger, were able to impose their rule on neighboring tribes. Above this real, materialized strength resulting from an easier digestive assimilation of food, there must be placed the imagined strength produced by the awareness of well-being, of inner satisfaction, and by the feeling of conscious pleasure. Cooked meat represents above all the overcoming of putrefaction. Together with the fermented drink it constitutes the principle of the banquet, that is to say the principle of primitive society."

Fire purifies our food

for eating,

and in agriculture fire

also purifies the field.

"This purification is truly conceived

as going deep in the earth.

Not only does the fire destroy the useless weed,

but it enriches the soil,"

is how Bachelard explains it.

Fire becomes a symbol

for anything that brings

a sharpening of awareness

or intensity of feeling.

We have said that it equates to the heat of any passion.

Bachelard makes direct connections of fire

with imagination, as well:

"Imagination works at the summit

of the mind like a flame."

The image of fire,

when a poet or artist uses it, brings together all the many ways we think of fire, all its opposing qualities. In Bachelard's words:

"Fire is,

among the makers of images,

the one that is most dialecticized.

It alone is subject and object."

I would say that not only is "fire" subject and object,

as in "The fire is hot" (subject)

or "in the fire" (object)

but also "fire" is a verb, as in:

"You're fired,"

although here the meaning may be a bit different.

This statement means you're out of a job.

Bachelard continues to discuss the many ways that we think of fire as the vitality of everything that is alive:

"What I recognize to be living—living in the immediate sense—is what I recognize as being hot."

"Heat is the *proof par excellence* of substantial richness and permanence: it alone gives an immediate meaning to vital necessity, to intensity of being. In comparison with the intensity of fire, how slack, inert, static, and aimless seem the other intensities that we perceive. They are not embodiments of growth. They do not fulfill their promise. They do not become active in a flame and a light which symbolize transcendence." The other elements

—earth, air, and water—

don't carry as much intensity as fire.

*H*uman complexes,

which cause us to react in certain repetitive ways,

are always holding in tension very different aspects

or qualities of personality.

Fire, especially, lends itself to making complexes,

because it naturally contains such fierce opposites.

Bachelard names two, the Promethean Complex,

which causes us always to want to exceed any authority

(to think we know more than anyone else on any subject)

and the Empedocles Complex.

The latter one happens when we are in a dreamy,
> thoughtful state by a peaceful fire:

"This very special and yet very general kind of reverie leads to a true complex in which are united the love and the respect for fire, the instinct for living and the instinct for dying,"

> is what Bachelard tells us.

In other words,
> we feel both drawn by the attraction of the fire,

at the same time feeling a respect for its force.
> Reflecting by a fireside brings us awareness

of the closeness of life that can be snapped
> and taken away suddenly.

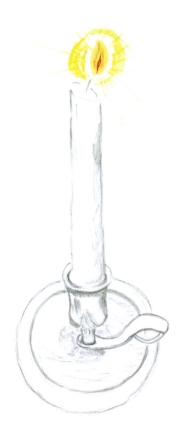

We may feel like falling asleep by a fireside, but the flame of a candle wakes us up.

Bachelard wrote a whole book on a single image, the image of the flame of a candle. The manner in which he approaches any image is to mention many of the ways in which it affects or works on us that we may never have thought about before. Then he follows with a discussion of what the image can teach us. He makes a good case that we learn much just by letting our imagination take in the simplest things of the world. He makes the example that we can learn from the flame of a candle to pick ourselves up and keep reaching to better ourselves. The flame teaches our will to stretch higher in the same upward and climbing way that a flame burns.

Bachelard
says
it this way:

"One who dreams
of the verticalizing will,
who learns the lesson
of the flame,
realizes that he
must right himself.
He rediscovers the
will to burn high,
to go with all his strength
to the summit of fervor."

Another of Bachelard's

favorite images

is the image of the phoenix.

He was writing about it

in a book that was never finished

when he died in 1938.

This mythical bird is a very ancient one.

Describing the phoenix, he said:

"The imagination is forced immediately
to come to terms with a fabulous being.

The Phoenix in fact is doubly the stuff of fable.

It both bursts into flame of its own fires,

and rises again from its own ash.

We, who no longer believe what our imaginations tell us,

must attempt to live this double miracle out.

Because people once believed
in the existence of the Phoenix,
it will be necessary for
us to believe in it a little ourselves,
to know it as it was once known."

Maybe some of you have seen Stravinsky's ballet of the *Firebird*. Bachelard catches the essence of the firebird:

"Our firebirds are not images of the substance of fire, but rather images of speed. Firebirds are flashes of fire."

Another bit of joyful,

flashing light is the luminous tail

of the firefly.

What would a summer's evening be

without fireflies in a jar?

At the opposite end of things, fire out of control

is certainly fierce. We have respect for it as being dangerous,

but whatever would we do without its benefits.

About the author

Joanne H. Stroud, Ph.D., taught both literature and psychology at the University of Dallas before becoming a Founding Fellow of the Dallas Institute of Humanities and Culture. Currently on the faculty there, she teaches in the Spiritual and Cultural Psychology program. She is also Director of Publications and Editor of the Gaston Bachelard Translation Series, which consists of seven works on elemental imagination. She lectures in Dallas, New York City, and Connecticut and writes frequently on Bachelard, a 20th century French Philosopher of Science. In 2002 in Dallas she chaired a symposium on Bachelard, "Matter, Dream, and Thought," that attracted international attention. Her book, *The Bonding of Will and Desire,* was published in 1994.

Tashasan practices illustration and obscurity in the south of England.

FIRE IS HOT

Copyright © 2005 Joanne H. Stroud
Illustrations Copyright © 2005 Tashasan

ISBN 0-911005-45-5

All rights reserved. No part of this publication may be reproduced, stored in any retrieval system, or transmitted in any form or by any means, mechanical, photocopying, recording or otherwise, without permission in writing from the author, except by a reviewer, who may quote brief passages in a review to be printed in a magazine or newspaper.

The Dallas Institute Publications publishes works concerned with the imaginative, mythic, and symbolic sources of culture.

Publication efforts are centered at:
The Dallas Institute of Humanities and Culture
2719 Routh Street, Dallas, Texas 75201
www.dallasinstitute.org

Printed and bound in China
with Palace Press International

Design by Suzanna L. Brown
Illustrations by Tashasan

First Printing, 2005

Air Is Clear

Air Is Clear

Joanne H. Stroud, Ph.D.

illustrations by Tashasan

Air
is what we need to breathe.

Just as water is the medium of life for fish,

air is the medium for humans.

We can't see it or taste it,

but we do feel it when it moves.

Air is light, although it can feel quite heavy when it is humid outside.

We say that something is "as light as air."

Air takes up space.

You can prove this by a little experiment

with a clear plastic cup,

a tissue, and

a bowl of

water.

What you do:

Roll the tissue into a ball and push it into the bottom of the cup. Turn the cup upside down and push it straight down into the bowl of water. Does the water fill the cup?

Lift the cup out of the water. Is the tissue wet?

ir does expand and contract.

As warm air expands it becomes lighter.

When the air in a hot air balloon is heated,

the inside air is lighter than the outside air,

causing the balloon to rise.

As warm air rises,

cold air has to fill the space below,

making a current or thermal.

Birds and gliders can get in the stream

of warm air thermals

and glide effortlessly for miles.

*A*ir in water creates bubbles.

Why do you think whales leap into the air?

Because it's fun.

And the resultant bubbles feel good too.

*A*ir improves the batter of bread or cakes.

When beaten, dough becomes much lighter.

Without air, soufflés and popovers would never

rise when they are baked.

Air creates bubbles

in a pot or kettle when water boils.

These rise to the top and escape as steam.

Air makes the kettle whistle.

Air is a colorless and tasteless gaseous mixture.

We think of it as being mostly oxygen, but actually it is made up of approximately 78% nitrogen and 21% oxygen. *Air contains small amounts of other gasses such as argon, carbon dioxide, neon, and helium.* When the atmosphere is stuffy, still, or without a breeze we say it is "airless," although that is never totally true. In some cities like Los Angeles and Mexico City, smog and haze are trapped by air that can't get over the mountains.

Air pollution then becomes a health hazard.

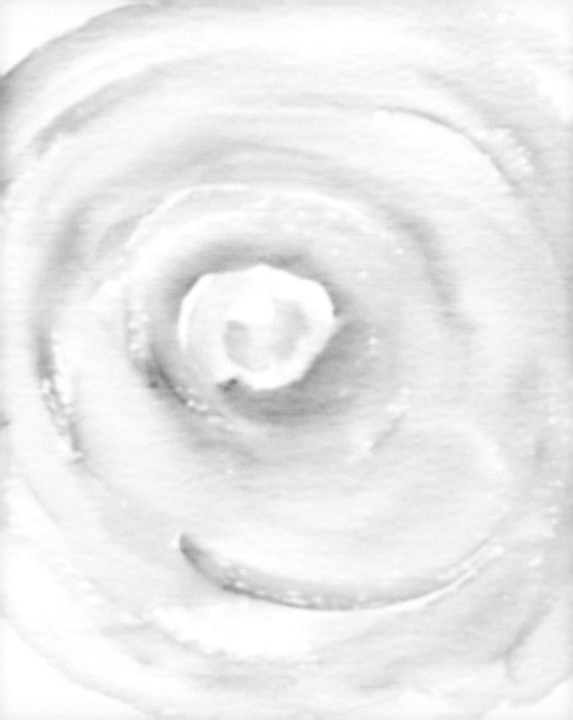

Air is our earth's atmosphere.

Atmospheric pressure is the pressure exerted by the atmosphere because of its gravitational attraction to earth. It is measured by barometers and usually expressed in units of inches (or millimeters) of mercury. The atmospheric pressure is greater at sea level than at the top of a mountain. In a place like Santa Fe, tennis balls fly faster due to the lighter air. Earth's favorable upper atmosphere protects us from lethal short-wave radiation from space.

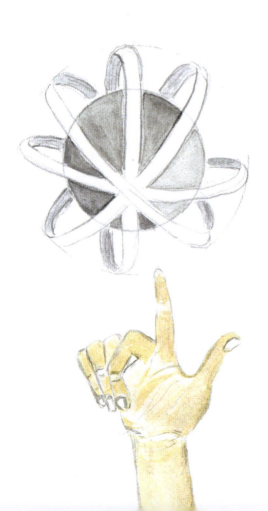

Air is wind when it is moving roughly parallel to the ground.

Windmills are still used in some parts of the world to capture the wind's energy. Four rotating blades attached to a wheel make it work. Windmills once were used to catch the wind and use its power to pump water or grind wheat into flour. Windmills can even make electricity from wind power. Today, too, an effort is being made to harness the wind with a whole series of windmills as an alternative form of energy.

Air flows.

Because the earth spins, upper winds generally flow from east to west in North America. However, strong wind-blown fronts bring changes, sometimes violent, in the weather. Polar winds from the north bring cold air, while south winds are usually warmer on our continent.

There are many devices that tell which direction the wind is coming from. Even small airfields have windsocks to show the direction and strength of the wind. Weathervanes on the rooftops of houses and barns tell whether it is a north, south, east, or west wind.

Air brings all the weather—breezes, wind, rain, thunderstorms, snow, sleet, blasts, gusts, gales, whirlwinds, whatever. We find the wind's music in mist, rain, and sea spume. Swirling air makes tornadoes, which we sometimes call "twisters," cyclones, typhoons, and hurricanes. It is called a hurricane when a tropical cyclone has winds that exceed 75 miles per hour. Hurricanes occur in the Atlantic Ocean or Caribbean Sea, and they generally travel northward. For some reason, hurricanes in the Western Pacific or China Sea are called typhoons, perhaps because they were named before anyone realized they were the same ferocious storm.

Air causes waves to break in the seas.

The height of the waves depends upon how high the wind is kicking them up. An on-shore breeze flattens the waves, while one coming from off shore stirs them up higher. *The New York Times* in a recent explanation made it clear that tsunamis follow the same laws of physics as ordinary surf waves created by wind. For wind-driven waves, the distance between wave crests—the wavelength—is at most a few hundred yards. For tsunamis, that wavelength can be a hundred miles or more. Because the wavelength is so much greater than the ocean depth, the speed of the wave depends on that depth. In water 2.5 miles deep, the average depth in the Pacific, a tsunami travels almost as fast as a jetliner, 440 miles per hour.

Air once long ago carried the seeds that changed the look of our planet.

During the closing of the Age of Reptiles, something over two hundred million years ago, simple primitive seeds that could carry some nourishment for a growing young plant developed. "After a long period of hesitant evolutionary groping, they exploded upon the world with revolutionary violence" is how Loren Eiseley puts it. He was a distinguished anthropologist, but often affectionately called a "poet-scientist."

By the bursting of its pod of seeds, "a plant, a fixed, rooted thing immobilized in a single spot, had devised a way of propelling its offspring across open space." They had not always been able to spread themselves this way. They had been restricted to one locale. "When the first flower bloomed on some raw upland late in the Dinosaur Age, it was wind-pollinated, just like its early pinecone relatives. It was a very inconspicuous flower because it had not yet evolved the idea of using the surer attraction of birds and insects to achieve the transportation of pollen. It sowed its own pollen and received the pollen of other flowers by the simple vagaries of the wind. Many plants in regions where insect life is scant still follow this principle today," Eiseley explains.

The seed contrasted to the earlier spore. The new seed at the heart of the flower grew independent of outside moisture. "The seed, unlike the developing spore, is already a fully equipped *embryonic plant* packed in a little enclosed box stuffed full of nutritious food. Moreover, by featherdown attachments, as in dandelion or milkweed seed, it can be wafted upward on gusts and ride the wind for miles; or with hooks it can cling to a bear's or a rabbit's hide; or like some of the berries, it can be covered with a juicy, attractive fruit to lure birds, pass undigested through their intestinal tracts, and be voided miles away."

Eiseley expresses it so beautifully that I can't improve on his words: "These fantastic little seeds skipping and hopping and flying about the woods and valleys brought with them an amazing adaptability.... All over the world, like hot corn in a popper, these incredible elaborations of the flowering plants kept exploding. In a movement that was almost instantaneous, geologically speaking, the angiosperms had taken over the world. Grass was beginning to cover the bare earth until today there are over six thousand species. All kinds of vines and bushes squirmed and writhed under new trees with flying seeds."

Air was a true helpmate in beautifying the world.

*A*ir as wind

        seems often to die down

                late in the day.

I've always liked the line depicting twilight in Thomas Gray's poem "Elegy in a Country Churchyard":

"Now fades the glimmering landscape on the sight,

        And all the air a solemn stillness holds."

Air makes it possible for birds to fly.

Thousands of birds in North America and Canada take to the skies every year in the fall when it is time for them to migrate south to warmer climates, returning in the spring. Birds can soar and dive straight down for fish. White, screaming seagulls circle when scavenging for food. Hummingbirds can hover over sweetened water by flapping their gossamer wings very rapidly.

Air is the medium of life for thousands of airborne insects too.

Many, such as butterflies, dragonflies, mayflies, and ladybugs, live a very short life. The cycle of the butterfly is particularly interesting. Starting out as a strictly creeping, land-based worm, the caterpillar, when it is time, spins a hard chrysalis around itself that gradually becomes clearer and clearer. Ten days to two weeks later a beautiful orange and black butterfly emerges. After a few hours of drying its wings, the butterfly takes to the air to find a mate and lay eggs to perpetuate itself. Butterflies in North America fly long distances to winter in enclaves in Mexico.

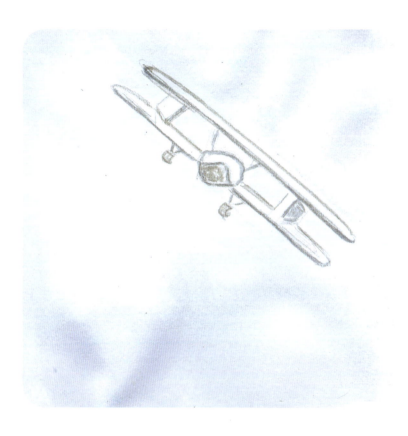

Air makes it possible

for people to fly.

Since the Wright brothers' invention,

people can fly in an airplane or helicopter.

It is quite surprising that by its design a heavy plane's wing,

with less airflow passing above than beneath,

has enough lift created in the pocket to climb high in the air.

When a plane drops altitude suddenly, it has hit an air pocket.

Hydrofoil boats glide on air on top of the water, and high-speed trains

use air to remove friction so they can travel at faster rates than

the traditional locomotive on rails.

Air carries a kite high in the sky.

Kites made their appearance over 2,500 years ago in China. In 1752 Benjamin Franklin used a kite in a storm to prove that lightning is a form of electricity. Every school child learns that Wilbur and Orville Wright are credited with the invention of the airplane in 1903. What most don't realize is that the Wright brothers were also skilled in kite flying. It was their years of kite flying that directly led to the invention of the airplane. A kite, an "aerial acrobat," is "looked upon in the Orient as a purifier of the atmosphere."

– Gaston Bachelard

Air caught in the sails of a boat
propels it ahead.

It is a thrilling sight to see a slanting schooner cutting through the water with two or three sails capturing the wind, or to see a sailboat gliding downwind with a flying spinnaker billowing in front. The poet Edward Hirsch captures the joyous lift exactly:

"A sail leaning out into the down drafts of the wind,
the anxious cross currents,
the up surges of the open air."

*A*ir is hard to see.

More often we see the effects of air—in leaves blowing, the rain falling down in diagonal sheets, clouds drifting overhead, or the way trees lean because of the prevailing winds. A friend of mine who is a fighter pilot tells me that one can actually see the air in little ripples flowing over the wing of a plane in the moment of passing through the sound barrier. When an aircraft breaks the sound barrier, we could even say that air creates sound.

*A*ir vibrates to create
all of the varying, melodious tones
of the different instruments in an orchestra.

This fact is obvious in the wind instruments such as flutes, clarinets,

trumpets, or harmonicas that use the breath, but it is equally true of violins,

violas, cellos, and basses where notes are created by drawing a bow

across a sound box with stretched strings.

Even all the drums depend upon the vibration

of air to create sound.

Airwaves carry sounds and electronic signals over vast distances.

Different mediums use various wave bands to transmit information.

Radios use radio frequencies, with high frequency in the range of

3 to 30 megacycles. Broadband and ultra high frequency bands are

also available. Radio waves traveling through the air open

garage doors and bring us wireless messages on our cell phones.

Air lends itself to many expressions.

Here are a few examples: An uncertain decision or event is "up in the air." When one feels elated, it's like "walking on air." When something seems widespread, it is "in the air." We say he (or she) "aired his (or her) opinion," when we are speaking about venting. "Aero" is a prefix for many words pertaining to air, such as aerodynamics, which means the interaction of air and objects, or aerosol, which is generally applied to a product packaged under the pressure of a gas. Aerodrome and aeroplane are words still in use in England. In the United States, we say airport and airplane.

Air as airiness is related to whatever is lively, jaunty.

Shakespeare in *A Midsummer Night's Dream* connects the lunatic, the lover, and the poet in one of the most lively descriptions of how our imagination works:

"The lunatic, the lover, and the poet,

Are of imagination all compact:

One sees more devil than vast hell can hold,

That is the madman:"

And then after talking about the lover, when he comes to the poet, he says:

> "The poet's eye, in a fine frenzy rolling
> Doth glance from heaven to earth, from earth to heaven
> And, as imagination bodies forth
> The form of things unknown, the poet's pen
> Turns them to shapes and gives to airy nothing
> A local habitation and a name."

This means that the poet can pluck out of the air an "airy nothing" and give it form and substance.

In Shakespeare's drama *The Tempest*, Ariel is one of the chief characters. He can fly through the air to do the bidding of Prospero, who has learned how to do magic. His opposite is Caliban, who is more nearly an animal and very earthy. At the end of the play, Prospero sets Ariel free. Actually, air and freedom have much in common.

In the Christian tradition

angels and archangels travel

through the air to visit the human race.

Some are pictured with beautiful, feathery wings

that are almost as long as the body.

Cherubs and fairies

also have small wings and can fly.

In the story of Peter Pan, Tinker Bell and Peter know how to fly. He teaches the Darling children how to use fairy dust to be able to reach make-believe Neverland. Harry Potter understands magic too. He and Ron borrow the family car that flies. Harry also uses a stick that zooms through the air to play very perilous cribbage games.

In the classic Greek tradition Iris was a messenger between immortals and humans.

A rainbow in the air was her staircase. Some say that she moved so quickly through the air in a dress of many colors that she left a trail behind. A rainbow was also Noah's sign from God that the floodwaters were going down.

We now think of mythology only as colorful stories. Actually there was a time when myths were firmly believed. It helped to have a deity to explain the mysteries of the universe. Myths still provide good images for understanding all the ways that any elemental aspect of nature behaves. Here is how a current writer describes myth in our time:

"Behind every myth

is a truth dying to get out,

a kernel of truth about the human condition

that needs an image."

- Phil Cousineau

Air in archaic Greece was personified by the God of the Winds, Aeolus.

Aeolus was usually drawn in pictures or depicted in statues with his cheeks full of air getting ready to blow it out. From his name we get the Aeolian harp that Keats mentions in his poem. Aeolus, given domination over the winds by Zeus, kept them trapped in deep caves on his island. If you remember the episode in the story of Odysseus in *The Odyssey*, it is Aeolus who gave Odysseus a bag of dangerous storm winds to keep them from blowing him off course as he sailed back home to Ithaca. For island peoples who needed good winds to propel their boats, having the god of winds on your side was vital. Odysseus' unruly sailors didn't trust him, and, thinking that he might have gold in the bag, opened it up, and all the storm winds escaped.

One of the most famous tales

of flying is the story of Icarus.

His father, Daedalus, fashioned wings out of feathers and wax for both of them to escape from the island of Crete where they had been imprisoned by King Minos. Daedalus had built the labyrinth confining the terrible Minotaur to whom Athens was required to send seven youths every year to placate. After taking off, Daedalus led the way, and all was well for the first part of the trip. Daedalus had warned Icarus not to fly too high or else the hot sun would melt the wax. When they were close to safety, Icarus forgot the warning and climbed upward before falling to his death, crashing into the sea while his horrified father watched.

The Athens Archaeological Museum has a headless

5th century B.C. statue called Aura.

According to the dictionary, "aura" is " an invisible

breath or emanation."

I think of aura as the light halo that some can see

around a person's body.

The stately Greek statue is described in a recent book as:

"a personification of breeze in the form of a woman whose

maker battled against his own material, doing all he could to

transform a block of stone that would crush bones if it fell,

into the illusion of a transparent current of air, the pleats of

her dress eddying unevenly around her, following the

topography of her body."

- Patricia Storace

Once again it is only by showing the effects of air

that we can view it. The famous statue of

the Winged Victory at Samothrace

also depicts in stone the effects of wind

against the thin material clinging to the body.

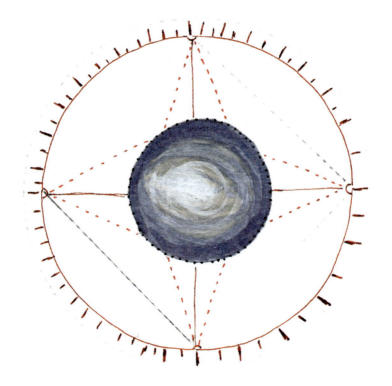

The four elements

—earth, air, fire, and water—

are each assigned to three of the twelve signs of the zodiac.

Air is the element for those born in Gemini, Libra,

and, surprisingly, Aquarius, who is the water bearer

and the 11th sign of the Zodiac.

Gaston Bachelard, the twentieth century French philosopher,

was fascinated with all the layers of meaning attached

to the four elements.

It is in *Air and Dreams* that Gaston Bachelard makes one of my favorite assertions about imagination. In it he is speaking about the continuous flow of images in the imagination and how they link together. He says:

> "I think I am justified in characterizing the four elements as the hormones of the imagination."

What I take this statement to mean is that each of the elements, when we let our minds dwell on them, starts a whole flow of other associated images. He follows with this:

> "Imaginary air, specifically, is the hormone that allows us to grow psychically."

Whatever does this mean?

He continues by explaining that thinking about air causes us to think about rising off the ground. So air tends to make us feel lighter and therefore better, even to feel elevated. It raises our sights and increases our hopes. It leads to "a healthy straightening up, growing tall, and carrying our heads high."

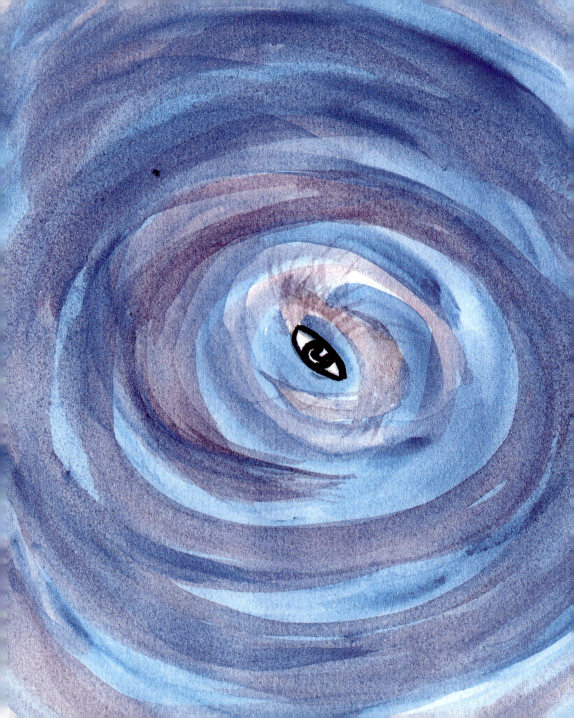

Images of air are difficult to pin down; they "either evaporate or crystallize," Bachelard remarks. He shows how the imagination works by a kind of quick flickering between reality and daydreaming. Just think of such words as wing and cloud. We find ourselves making pictures of them, but then in another instant the imagination rushes to add new clouds, wings, and other imaginative thoughts. This happens because the imagination is naturally dynamic, taking us beyond and freeing us from only what the senses perceive.

*I*n a chapter on

"The Poetics of Wings" Bachelard says:

"The sky is a world of wings."

Bachelard is an acute observer of the natural world that is all around, everywhere. Even though it is directly in front of us, how often do we ever stop really to see? He thinks of everything in a new light and with lively connections in such statements as this:

"The wing is essentially aerial.

We swim in the air, but we do not fly in the water."

$\mathcal{B}$achelard, our guide to the way images unfold, then goes on with remarks about how birds have always stirred the imagination of all those who love the natural world and the beauty of flight. Religious leaders such as St. Francis, and poets, particularly the English Romantic poets, have sung about this way of raising our thoughts in a heavenly direction. Bachelard describes the joy of some individual birds that have attracted poets: "The startling invisibility of the lark has never been sung better—as a wave of joy—than by Shelley in "To a Skylark." Shelley understood that it was a cosmic joy, an 'unbodied joy,' a joy that is always new in its revelation."

What this means to me is that it is not only when we see the beauty of the flight of the bird but even when we only hear its joyful trill that our hearts are lifted. Shelley sees the skylark as a messenger of rapture:

> "Teach us, Sprite or Bird,
> What sweet thoughts are thine:
> I have never heard
> Praise of love or wine
> That panted forth a flood of rapture so divine."

In like manner, John Keats in his "Ode to a Nightingale"

also sings the praises of the song of the invisible nightingale:

"Hail to thee blithe Spirit

Bird that never wert."

The later nineteenth century poet Gerald Manley Hopkins in "Windhover" writes a whole poem about the gladness in his heart on seeing in the early morning the flight of a bird:

"My heart in hiding

stirred for a bird,—the achieve of, the mastery of the thing!"

Even the little lark is a messenger of joy:

"The lark carries the earth's joys to Heaven,"

the French poet Michelet claims. Look at this reverse order:

not that the bird brings the joys of Heaven to earth

but the other way around. Perhaps a bird just

sings as a way of breathing.

Finally, returning to Bachelard

who reminds us

that birds, sky, freedom—Air—all

inspire us to rise to greater heights.

What better place to end.

I'm glad you're there,

Air.

About the author

Joanne H. Stroud, Ph.D., taught both literature and psychology at the University of Dallas before becoming a Founding Fellow of the Dallas Institute of Humanities and Culture. Currently on the faculty there, she teaches in the Spiritual and Cultural Psychology program. She is also Director of Publications and Editor of the Gaston Bachelard Translation Series, which consists of seven works on elemental imagination. She lectures in Dallas, New York City, and Connecticut and writes frequently on Bachelard, a 20th century French Philosopher of Science. In 2002 in Dallas she chaired a symposium on Bachelard, "Matter, Dream, and Thought," that attracted international attention. Her book, *The Bonding of Will and Desire,* was published in 1994.

Tashasan practices illustration and obscurity in the south of England.

Air Is Clear

Copyright © 2005 Joanne H. Stroud
Illustrations Copyright © 2005 Tashasan
ISBN 0-911005-44-7

All rights reserved. No part of this publication may be reproduced, stored in any retrieval system, or transmitted in any form or by any means, mechanical, photocopying, recording or otherwise, without permission in writing from the author, except by a reviewer, who may quote brief passages in a review to be printed in a magazine or newspaper.

The Dallas Institute Publications publishes works concerned with the imaginative, mythic, and symbolic sources of culture.

Publication efforts are centered at:
The Dallas Institute of Humanities and Culture
2719 Routh Street, Dallas, Texas 75201
www.dallasinstitute.org

Printed and bound in China
with Palace Press International

Design by Suzanna L. Brown
Illustrations by Tashasan

First Printing, 2005

Earth is Round

EARTH IS ROUND

Joanne H. Stroud, Ph.D.

illustrations by Tashasan

Earth is round.

When we look

at a globe or see it from above,

it is certainly round. Now we know that to be true.

But it was hard to convince some of the earliest peoples,

because walking along it looked quite flat.

Even now we speak of the four corners

of the earth as if it were a flat map.

Earth is

enormous.

Just to hold the idea of it in

our mind's eye

causes our imagination to grow and expand.

The word "vast" awakens the idea of a larger universe.

Our imagination loves to be filled with anything

that is too big for it.

EARTH

IS HOME.

EARTH IS HOME FOR MANY CREATURES OTHER THAN THE HUMAN RACE.

Wislana Szymborska,
1996 Nobel Prize winner,
says it like this:

"A miracle,

just take a look

around:

the inescapable earth."

EARTH

IS BOTH THE GROUND UNDER

OUR FEET AND THE WHOLE WORLD,

THE PLANET.

Therefore, earth is both

very near at hand and,

at the same time,

what we would see from afar

if we were looking from an astronaut's

window orbiting in space.

Generally, we call the first simply "earth,"

but the second "<u>the</u> earth."

Earth is matter and mother. The word for "matter" is related to the word for mother in many languages. That's why we talk about mother earth.

Earth

nourishes us.

"The earth that feeds us all,"

is how Homer calls her.

We are thankful for vegetables that grow

from the richness of her soil and the

abundance of trees with their fruits and nuts.

Many parts of the world that once were full of trees are now

deserts. Laying waste to valuable trees is currently

a big issue—in the rainforest of the Amazon,

for instance, because too many trees are chopped down

to use for fuel or to make room for development.

Undiscovered cures for diseases may be lost when

primal forests are destroyed.

Earth needs water to grow most plants.

Even the desert comes alive with flowers when irrigation brings water to it. Earth can be a richly colored brown mud when mixed with water, as everyone knows who has ever made a mud pie. Earth can turn into a swamp with too much water. Or earth can turn into dust without water. Someone said:

"When you pick up a piece of dust, the entire world comes with it."

Earth sprouts a carpet of wild flowers.

Seeds from plants and nectar from flowers

support a whole cycle of living beings.

The many varieties of fungi sprouting

from the earth

(what we loosely call mushrooms)

play an important role

in our ecology.

Even the lowly earthworms

in the ground break

up the soil.

And insects have

their place in the life

cycle as well.

Earth
supports
animal life.

All

living creatures

depend upon the soil of the earth for life.

We forget that the birds in the air

and the fish swimming in the seas

depend upon earth's bounty.

Earth's animals

                are numerous.

Some no longer wander the world.

Dinosaurs are now extinct. Some think an asteroid

crashed into the earth and the resulting cloud cover

destroyed them. Buffaloes, too, once thundered the western

plains in large numbers but are now reduced to such few

numbers that they are just a curiosity.

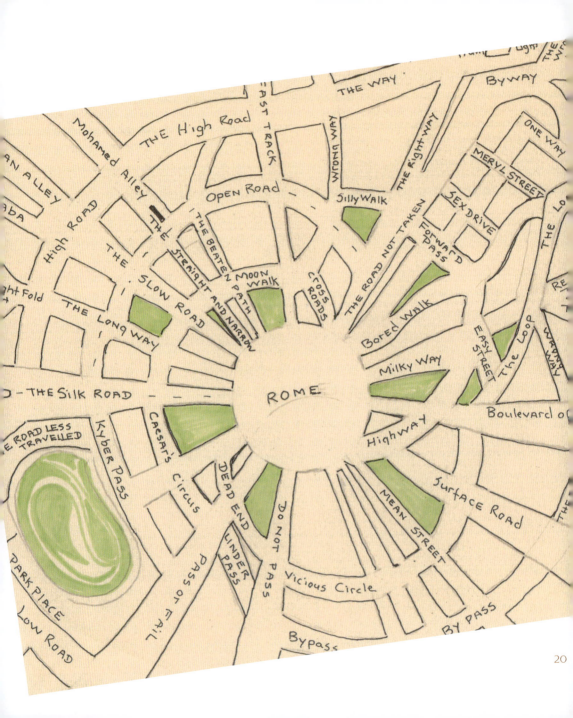

Earth was called by the Greeks, Gaia, or Gē, from which we get our word "geography."

From this root, we also get our word "geometry," which comes from geō-metria, or land measurement. According to Hesiod in the Greek creation myth, Gaia was the first creature born out of Chaos, which is how it all got started. Gaia mated with Ouranos, the sky or heaven god, to produce the Titans, the first generation of gods and goddesses. An ancient Greek fragment recognizes the earthly/heavenly connection, proclaiming:

"I am a child of earth

and the starry heavens."

Earth's name was

Prithivi

to the Hindus

and Ishtar

to the Babylonians.

In Egypt, the mother/father figures were opposite

in their gender to those in most other cultures.

There, the masculine figure was Geb,

the earth god, who mated with Nut,

the sky goddess.

Earth and sky have always

spoken to each other.

Rabindranath Tagore,
the great Indian poet, calls

trees -

"eternal attempts by the earth to

speak to the listening sky."

Earth is a place
of birth and a place of burial.

In our time burial in earth is a common practice. The ancient Egyptians constructed huge pyramids for their pharaohs to go into the journey of afterlife. They wanted the ones they loved to have a comfortable travel to the next world, so the inside chamber of pyramids held articles that were important to the dead. The earliest Christians were buried in underground catacombs with recesses for graves.

Earth

has layers

or regions,

each quite different.

There's the upper world of the gods

and the there's the underworld, where ancients

believed everyone went when they died.

Ophalos is the name given to the point that ancient

Greeks used to mark the center of the earth.

As very small children many of us believed that if

we could dig deep enough, we would come out on

the other side of the earth.

Earth is Demeter's realm

in Greek mythology,

and so she spends little time high above on Mount Olympus with the other Olympian gods and goddesses. Demeter's daughter, Persephone, was snatched by the god of the underworld, Hades, and dragged underground. She had been picking flowers when Hades fashioned a glorious narcissus to lure her, a narcissus that the *Homeric Hymn to Demeter* (5-15) says carried a fragrance that

"made all the wide sky and earth to smile."

Demeter,

goddess of grain

and plant life,

was

so distraught that she denied

growth to all green things until Persephone

was allowed to return to earth for the larger part

of each year. Persephone's status was permanently

altered. She had tasted the fruit of the underworld, the

pomegranate seed, and because of this trap, Hades demanded

that she remain part time as his Queen of the Underworld.

These are the three winter months when all green

things remain buried underground.

Earth holds the remembrance

of spilt blood and the urge for revenge. The Greek dramatist Aeschylus wrote an entire play all about the Furies, who were bitter, vindictive forces of the earth. This play, *The Eumenides*, is the last one of the only complete trilogy that has survived from the classical Greek period. It was produced in Athens in 485 B.C. It is a very complicated tale of hideous murders and the terrifying revenge that followed between two brothers who were rulers of the House of Atreus.

Here is a brief summary:

Already in the previous generation this family had engaged in murderous killing of close kin. Unforgiving hate reigned. One son, Agamemnon, who was the leader of the Greeks in the Trojan War, returned home with his concubine Cassandra, the sister of Paris and Hector. Agamemnon's wife, Clytemnestra, killed him for his cruelty in sacrificing their daughter in order to go to war. Orestes, their son, was in a terrible dilemma.

*U*nder the old rules, he had to kill his father's murderer, who was also his mother. He most reluctantly accepted his role. For this terrible deed, he was haunted and chased mercilessly for many years by the Furies, who always rise up from the earth to object when blood is spilt between close relatives. Orestes was finally brought to trial. The jury was a hung one, because half the jurors believed that Orestes was in an impossible bind between his loyalties to his father and to his mother and had suffered enough punishment to be forgiven.

The Goddess Athena

broke the tie and voted for his release. She persuaded the Furies to soften and forgive Orestes, and to join her beloved city of Athens. Henceforth, they became not the Furies living underground but transformed and supportive, residing at the base of the Acropolis with a new name, the Eumenides, or kindly ones.

EARTH

GIVES US

A SENSE OF

COMMUNITY.

IT REMINDS ME OF COMMON INTERESTS

AS HUMAN BEINGS

BECAUSE

IT BELONGS

TO ALL OF US

WITHOUT DIVISION.

*Earth beckons us
to get to know our surroundings.*

We are inspired and challenged to make something from

this meeting—to fashion a clay pot to carry water or

to tame a horse to carry us. In our day, in order to

make the world seem smaller, we have

produced wireless communication.

Earth provides us with the matter

or materials to build and construct. Wood for struts and beams, metals for girders, all are substances or the stuff of the earth. From this encounter we have developed the professions of engineers, architects, builders, and landscape architects. Then real estate agents, painters, plumbers, electricians, and gardeners follow to oversee what we build.

Earth is

not inert
or dead.

Earth has energy lines and flow.

Siting of temples at the points of energy

was the practice in early Greek times.

And Native Americans always pick these highly charged

spots for ceremonies of worship.

Eye of the Needle Island
The Edge of the Known World
Schizophrenia
The Razor's Edge
The Void
Point of No Return
The Great Schism
Timbuktu
Lapland
Tinker's dam
Gothland
Way Out There
Hither
Yon
Back of Beyond
Beyond
Yonder
Backwaters
The Pale
Beyond the Pale
The Brink
The Inky Deep
Land's End
Kamiza
Kishmer
Middle Kingdom
Kunlun
fro
to
Now
Sphere of Reality
Here
Hop
Limbo
Skip
Sump
Down Under
the Outback
The Underworld

Outer Mongolia
The Great Beyond
The Land that Time Forgot
Chaos
Isles of Langerhans
Dreamland
The Land of Nod
End of the Rainbow
Utgard
Valhalla
Nirvana
Camelot
Absentia
Cloud Nine
Seventh Heaven
Fantasyland
Oz
Shangrila
Narnia
The Web
Xanadu
Wonderland
Altered Reality
Byzantium
Neverland
Lalaland
No Man's Land
Bermuda Triangle
Nether Regions
The Dark Continent
Barbaria
Wasteland
Scylla
Hinterland
Badlands
Charybdis
Point Blank
Dunce Steppes
The Straits of Narrow
Bolivious
The Pits
Round the Bend

Earth

inspires

an attitude of

wonder

and of awe.

Haven't all of us been stunned by the beauty of an early sunrise or the setting sun? The magnificence of it all! The English poet William Wordsworth captures this sudden joyful feeling in a poem:

"The earth and every common sight,

To me did seem

Appareled in celestial light,

The glory and freshness of a dream."

-"Intimations of Mortality"

Earth

responds

to our grateful

wonder

of it.

The moment we thank the stream for flowing, it becomes more alive for us. If we feel closer to the earth, we feel more care for trees and for every growing thing. The famous artist Michelangelo was inspired by the beauty of the earth when he said:

"My soul can find no

staircase to Heaven

unless it be through

earth's loveliness."

Earth

is not only lovely, but it is also a resisting element that turns out to be a great teacher. It has both hard rocks and soft sand to tempt us or thwart us. Earth, in its hard forms, as rock or mountain, inspires us to try to conquer it. That's why there are always teams of people attempting to climb the highest mountains such as Mount Everest in Nepal. Two of the most famous rocks are the Rock of Gibraltar and the rock that Sisyphus, in mythology, as punishment had to push up a hill every day. This gigantic boulder rolled back down every night in Sisyphus's futile attempt to reach the summit. His name is often used to express any activity that seems never to go anywhere.

Earth's

surface is not at all

like it is a few miles down.

There we find the hot lava that pours out of volcanoes.
Islands are continually being formed in the sea from the out-
flow of lava. Water is found if we drill a well in the right place.
In many parts of the world, water is so needed that the dousing
method is used. By drilling downward into sand and rock, gas
and oil are released, which supplies us with so much energy.
Layers of peat that grows in the bog supply fuel for
fires in places like Ireland. Metals in the earth
demand that we find ways to
penetrate the surface.

The lure of gold, silver, diamonds, and other jewels spurs us to dig deep in the earth's crust. Just thinking of a diamond can make our day sparkle.

Earth

can be

terrifying.

When the earth rumbles and shakes in an earthquake, it can be quite destructive and kill many people. This happens when one plate moves in one direction and another in the opposite direction, and the earth splits open.

Then the "terra," the word for earth in Latin, is not FIRMA at all. From terra, we get our word for "territory."

Earth is abundant
in it's variety—

mountains, forests, ridges, valleys, deserts, lakes, and oceans. Think of them all—tundra, permafrost, and icepacks near the north and south poles, barrier reefs, and cliffs, even under the sea. Above ground, majestic mountains capture snow in winter, while low valleys trap water in lakes. Fully seventy-five percent of our planet is covered with water, most of it salt water.

Earth's

surface is gouged in the most engaging

variety. Water carves crevices in the earth's

surface—deep ones such as the Grand Canyon. Caves in

the southern part of France, like Lascaux, have paintings of

bison and antelope from the Ice Age, when a reverence was

felt for the animal killed. Other famous caves are the one

described by Homer on the island of Ithaca and Porphrys'

Cave of the Nymphs, which is called the

Womb of the Earth.

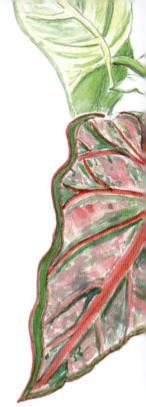

Earth

has a variety of different climates.

Generally, it is warmer the closer one gets to

the equator and colder near the poles. However,

Mount Kilimanjaro, in Africa on the border between

Kenya and Tanzania and very near the equator,

has snow at its top all year.

Earth also has a variety of seasons.

In the temperate zones, winter, spring, summer, and fall have quite distinct qualities with leaves on trees that fall in the autumn and come back to life in the spring. As one gets nearer the equator, seasons are less pronounced with hot weather present all year. Due to the tilt of the earth's axis, north and south of the equator the seasons are reversed.

Earth

has been divided up on maps into

segments called latitude and longitude.

This helps with navigation by boat or

by plane. Now we have man-made satellites

circling the earth that give us

global positioning,

even in our automobiles.

Earth has a natural satellite,

the moon,

in the same orbit around the sun.

So much that happens on earth

depends on the moon

that moves across the sky in a leisurely fashion.

The English word "moon" comes from

the Old High German word mano,

which means

"wanderer in the sky."

Earth, moon, and sun
        are at different angles
                to one another
           during their cycles.

The full moon appears in all its glory every twenty-nine and a half days, only to disappear entirely two weeks later. One half of the moon is always lit by the sun, but this lit side isn't always facing the earth. At the time of the new moon, the moon's sunlit half is turned away from the earth. At the full moon, the moon's sunlit half is fully facing the Earth. The moon is waxing when it is getting larger on the left side. Then it can be seen only in the early evening sky before it sets. The full moon rises at sundown and can be seen throughout the night before it sets at sunrise. The waning moon, in its last quarter, rises at midnight and sets at midday. Our word for month relates to the word "moon."

EARTH'S GRAVITY AND OUR MOON'S GRAVITY

ATTRACT ONE ANOTHER.

The tides of the oceans are one

of the most obvious results.

In every twelve-four period,

there are two high tides and two low ones.

Shellfish in the sea react to the moon

and the tides.

For example, lobster shed their shells

according to the moon cycles.

And weather is also related.

Earth has a strong

magnetic field centering

on the North Pole.

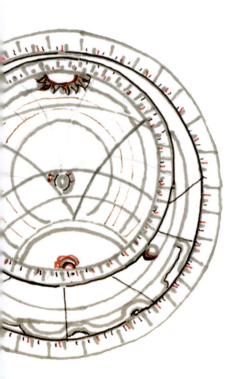

This attracts the magnetized needle of a compass, giving us directions. Magnetic fields are fascinating and are used in many ways in science and industry. You can see how powerful magnetism is by seeing the startling way that loose iron filings jump in semi-circular patterns when exposed to the ends of a magnet.

Gaston Bachelard, a French philosopher, in
the middle of the twentieth century began to talk about the
elements in a very special way. Earth is one of the basic four.
He wrote two books to describe earth—one about the active
side of earth as an active challenger

and

the

other about earth as more passively
accepting. The additional elements—water, fire, and
air—also required several books to explain them in
modern language. Of course Bachelard wasn't the first
to divide the essential components of the
physical world this way.

Starting with Plato and, later in the Middle Ages, and still later in the Renaissance, philosophers of these different ages have made this four-fold division. The

Eastern approach in contrast includes five elements—earth, water, fire, wood, and metal.

What is different about Bachelard's approach is the way he expands the imagination around each of the four elements.

Bachelard
talks in his book *Earth and Reveries of Will*
of the many ways we use our will. We may have
never before associated earth with the will, but Bachelard
explains how the earth conditions our will. We use it not just to
get a job done, to see a project through to completion, what we
might call determination, but in many other ways as well. Bachelard
also speaks of the "dreaming will," which is connected directly to our
imagination, or the ability to make images. He often talks of "reveries,"
which are not just night-time dreams while we are sleeping. He calls our
attention to the power of daytime dreaming, which he claims is a
treasure house of inspirations for the imagination. Others have
somehow made day-dreaming sound like a waste of time.
Not Bachelard! He shows that all
our good ideas and plans begin in images formed
when we are quiet and receptive.

Bachelard reminds us of the myth of Atlas, who is said to have held the world up on his mighty shoulders.

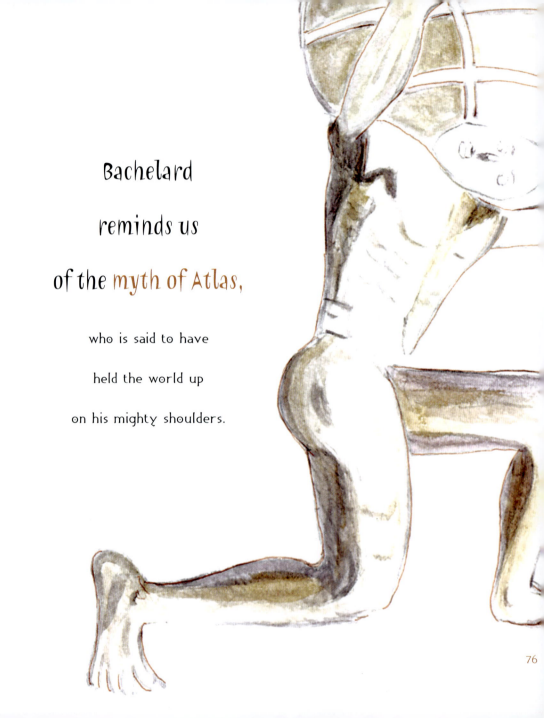

The hero Hercules, in one of his tests of strength, comes along and offers to share the burden, but gives it back to Atlas. To hold this story in our imaginations for a while causes us to feel stronger ourselves. Our imagination is naturally expansive. When the energy of the imagination is sensed, we begin to feel more and more powerful, and finally all-powerful. "Reveries of the will to power are reveries to be all-powerful," is how Bachelard explains it. That said, it follows that to imagine effort in this way makes the whole body strong, not just some muscles prevailing over others, as happens in purely physical exercise.

Bachelard praises the value of work,
especially with the hands or the body.

work is most satisfactory when we accomplish it ourselves. Bachelard says this is true because: "the will to work cannot be delegated. It derives no benefit from the work of others. The will to work prefers to do." wendell Berry's poem says much the same about the joy of actively working the earth:

"Sowing the seed,

my hand is one with the earth.

wanting the seed to grow,

my mind is one with the light.

Hoeing the crop,

my hands are one with the rain.

Having cared for the plants,

my mind is one with the air.

Hungry and trusting,

my mind is one with the earth.

Eating the fruit,

my body is one with the earth."

— wendell berry

Bachelard speaks

of the different tools

that humankind has used through the ages as a way

of responding to the world. He talks about the big

advance in the knowledge of good when earliest

humans started using the hammer instead of the club.

A simple rock clutched in the hand was the first

weapon of cruelty, or as Bachelard reminds us,

"Rocks teach us the language of hardness."

As this evolved into a stone chisel, mankind began to enjoy the products that were possible. A stone hammer, made by joining a rock to a handle, extended the possibilities. As a result, indirect thoughts and ideas "stirred to life in the human brain, intelligence and courage formulating together an application of energy," Bachelard elaborates. He adds: "Work—work against matter—is the immediate benefit." It is in our work to master the physical world that we make our mark as individuals even today.

Each element

has the ability

to engage us,

but earth especially

challenges our curiosity

and imagination.

Bachelard

uses this example to explain how the physical nature of a simple aspect of the earth teaches us a lesson in how to conquer whatever opposes us. For example, granite does not break easily or scratch or fall apart, and therefore it serves as an example of stability: "The very texture of granite gives expression to the permanence of being."

Mud and clay have been important over the historical period of human life.

Bachelard draws many examples of the possible levels of wetness of dough or paste and the hand's enjoyment of fashioning a vessel or, with flour, a cake. But it is not only the hand that finds satisfaction in mud. When the weather is warm, we are always tempted to take off our shoes and feel the sand or grassy earth against our feet.

Bachelard writes about this experience:

"walking barefoot in...mud awakens our own primitive, natural connections with earth."

In the volume *Earth and Reveries of Repose,*

Bachelard talks about places of refuge.

Here we follow him into the closed spaces on and of earth—the house, the cave, the grotto, even the labyrinth. The most famous labyrinth was that of the Minotaur on the island of Crete. The Athenian hero Theseus sought to end the sacrifice of seven youths and seven maidens that were sent each year to satisfy the Minotaur of King Minos. Theseus slayed the Minotaur and, with the help of King Minos' daughter Ariadne, half-sister of the Minotaur, escaped from the labyrinth. She gave him a thread to find his way out from deep in the earth. When Theseus abandoned Ariadne, she became the bride of the god Dionysus.

Besides caves,

roots are another

image of earth

that Bachelard

explores.

In mythology,

roots, vines, and wine

are the gifts

that the god

Dionysus, or Bacchus

as the Romans

called him,

gives us.

coming

round and round we

are filled with love in our

hearts for our earth.

finally, one can't improve on the poet
Robert Frost's words:

"Earth's the right place for love.

I don't know where

it's likely to go

better."

About the author

Joanne H. Stroud, Ph.D., taught both literature and psychology at the University of Dallas before becoming a Founding Fellow of the Dallas Institute of Humanities and Culture. Currently on the faculty there, she teaches in the Spiritual and Cultural Psychology program. She is also Director of Publications and Editor of the Gaston Bachelard Translation Series, which consists of seven works on elemental imagination. She lectures in Dallas, New York City, and Connecticut and writes frequently on Bachelard, a 20th century French Philosopher of Science. In 2002 in Dallas she chaired a symposium on Bachelard, "Matter, Dream, and Thought," that attracted international attention. Her book, *The Bonding of Will and Desire,* was published in 1994.

Tashasan practices illustration and obscurity in the south of England.

Earth Is Round

Copyright © 2005 Joanne H. Stroud
Illustrations Copyright © 2005 Tashasan
ISBN 0-911005-43-9

All rights reserved. No part of this publication may be reproduced, stored in any retrieval system, or transmitted in any form or by any means, mechanical, photocopying, recording or otherwise, without permission in writing from the author, except by a reviewer, who may quote brief passages in a review to be printed in a magazine or newspaper.

The Dallas Institute Publications publishes works concerned with the imaginative, mythic, and symbolic sources of culture.

Publication efforts are centered at:
The Dallas Institute of Humanities and Culture
2719 Routh Street, Dallas, Texas 75201
www.dallasinstitute.org

Printed and bound in China
with Palace Press International

Design by Suzanna L. Brown
Illustrations by Tashasan

First Printing, 2005

Water Is Wet

Joanne H. Stroud, Ph.D.

illustrations by Tashasan

That seems pretty obvious.

But what else is water? After the first describing word,

"wet," comes to us, let us try to expand our thinking and

consider all the other ways that water relates to us.

Water moves.

Think of a flowing river or the waves in the ocean.

Water is stirred up by the wind or by the contours

of the earth or by currents in the seas.

The Gulf Stream is a good example,

with its strong current or river of warm water

flowing out of the Gulf of Mexico in the south,

then upward along the east coast of the United States

and all the way into the north Atlantic Ocean,

keeping those waters warmer

than they would otherwise be.

Water does

        not like to be static.

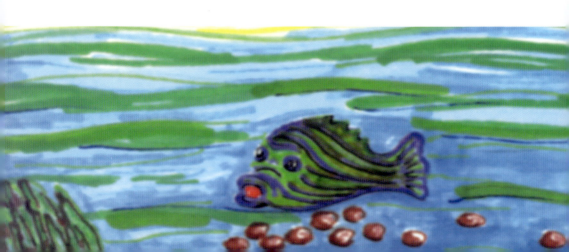

It swirls in decided patterns even when

it appears to be going straight.

　　　　Fish, whose natural habitat is water as ours is air,

　　know where to find the water

　　　　that has the right amount of swiftness of current,

　　　　　　where they can both eat and rest.

Water doesn't
           flow in
  a straight line.

Left on its own, it meanders.

Water that is constricted in a concrete ditch loses its living

quality for this reason.

Water reflects.

When it is still for a moment,

we can use it as a mirror.

Probably the first time that earliest humans saw themselves

was in a still lake or a container holding water.

Water mirrors the sky.

On a cloud-covered day,

the oceans are a somber gray.

On bright sunny days,

the oceans are a sparkling blue.

And in the tropics,

with a sandy-bottomed floor,

the oceans can be brilliant aquamarine.

Water itself
is many-colored,
not one
single color.

It can be transparent and clear,

or it can take on many different hues,

from dirty brown to bright emerald.

Water has

no taste

and no odor.

If it does,

something

has been added.

Water has a surface tension.

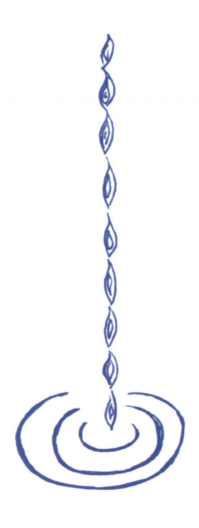

This quality helps hold up a piece of wood

or our bodies when we float.

Isn't it remarkable that thousands of people can ride

over the waves in a ship?

Water provided one of the earliest modes of getting around the world.

Since travel was easier by river or sea,

major centers where people collected were usually on waterways.

Rivers often mark the boundaries between countries.

T. S. Eliot calls attention to the nearness of water in our lives:

"The river is within us,

the sea is all about us."

(The Dry Salvages)

Water is the very substance of existence.

Our bodies are more watery than solid.

The Koran says:

"**By means of water,**

we give life to everything." (21:30)

Water is a good conductor

of heat, cold,

and electricity.

And our bodies, being made up of so much water

are also fairly good conductors.

For this reason, during thunderstorms with lightning in the sky,

a swimming pool or golf course is not the best place to be.

Water is

what we must have

to live.

Hard as it is to go without food, it is harder still to go without water. A person can go without food for many days, even weeks. Think of the fasts that Gandhi and others have undertaken to bring the public's attention to an issue. A person who has more fat will live longer than one who has very little. The body will use its fat and protein stored in the muscles to help it survive. So how long a person survives varies a lot. This is one case where one can be too thin. On the other hand, without water to drink a person will die within 3 or 4 days. It's the same for all sizes of people. Plants and animals also must have water, although the amount needed varies. Cactus plants can get along with very little water. Some plants, like water lilies, live in water. From water, trees get their sap.

Water is

one of nature's

greatest gifts

to us.

Because it is a necessity for all living beings,

many sages remind us to feel grateful for its abundance.

A good example is Benjamin Franklin, who said:

"When the well is dry, we know the worth of water."

- Poor Richard's Almanac, 1746

Or the ancient Chinese Proverb:

"When you drink the water,

remember the spring."

Water mixes easily

with a number of

other liquids

and solids.

We use it to dilute the intensity of many substances.

Water mixes with sand in a paste or with clay,

the material used in the first vessels that early humankind made.

Water also mixes and thins the batter for bread or cakes

to give it just the right amount of stickiness.

Water also corrodes.

Rust is made when water wears away

or coats metal.

Water's chemical name is H_2O.

That means it has two molecules of hydrogen and one of oxygen.

Water is healing.

We are told to drink 8 glasses each day.

We wash our wounds with water.

We also know how healing hot water is when we have

aches and pains.

And doesn't that first splash of cold water on the face

that wakes us up in the morning feel good?

Water—

when we drink water, it needs to be fairly pure.

Disease and plagues can spread rapidly

through a contaminated water source.

Water evaporates easily

in the desert or

dry climates.

The Dead Sea is really a large lake between Israel and Jordan.

It probably was once joined with the Mediterranean

but became salty by being land-bound.

Now it is 1,302 feet below sea level.

Even with a heavy inflow of water from the River Jordan,

its waters evaporate quickly,

leaving it even more salty than the ocean.

While the oceans have 4% to 6% of salts,

the Dead Sea has 23% to 25%.

Water makes up
most of
our planet.

That the land mass is less than 30%

is obvious when viewed from space.

The Rover's prod on the planet Mars is looking

mainly for signs of where water may have once been.

To sustain life on an alien planet

would depend upon a source of water.

Water takes

many

different forms.

Depending upon its temperature,

water can be quite a different substance,

from a gas to a liquid to a hard solid.

When over 100 degrees centigrade or 212 degrees Fahrenheit,

it burns as steam.

When below 0 degrees centigrade or 32 degrees Fahrenheit,

it freezes into solid ice.

Then it can be strong enough to skate on.

Water falls

  downhill

in a waterfall or

  in a raindrop.

When raindrops freeze, they bombard us with hail.

This happens when raindrops within storm clouds get blown

upwards and freeze. When tossed up and down like bouncing balls,

raindrops add layers of ice until heavy enough to fall.

Occasionally, they grow as large as grapefruits.

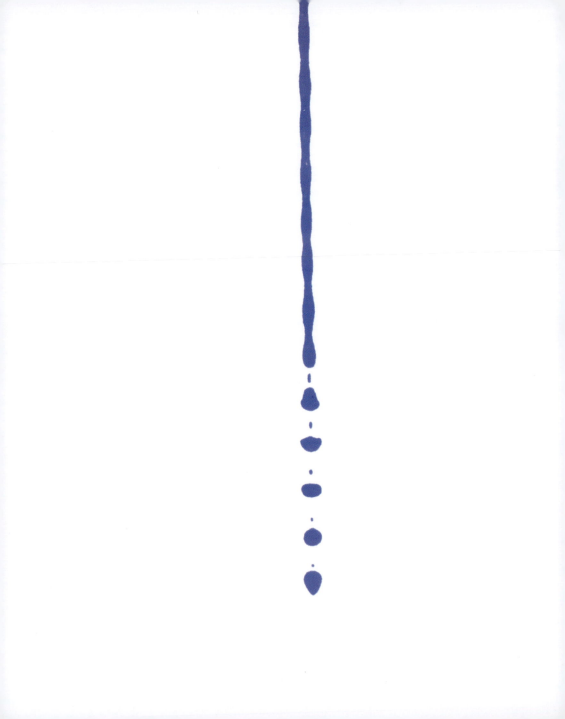

Water,

when it drips,

turns into a

series of

elongated

spheres.

Water

also falls

out

of the sky

as

snow

in winter.

Snowflakes are each a beautiful,

individual, six-sided crystal

if we look at them under a magnifying glass or microscope.

Crystals actually take many different shapes depending

on the temperature. From 27-32 degrees Fahrenheit or 13-18 degrees

Centigrade, they are shaped like hexagonal plates jutting in

all directions. Air closer to 32 degrees makes them turn into needles.

At temperatures between 3 and 10 degrees,

they become the familiar flat,

lacy patterns we see on a dark surface.

These are called dendrites.

Once again, we see how much more complex water

is than we thought.

Water vaporizes

on the ground

in the

morning dew.

Glistening on leaves and grass

in the cool of the early day before the sun dries them out,

dew sparkles like diamonds.

A spider's web, a thing of wonder in its pattern,

becomes totally beautiful with beads of water dripping

from each looping thread.

Fog is condensed water vapor that hugs the ground

in cloud-like clumps when the ground and air temperature vary.

Fog can make driving or flying difficult.

Fog can even have a color.

T. S. Eliot, though American-born, lived in London and wrote of

"The yellow fog that rubs its back upon the window-panes."

("The Love Song of J. Alfred Prufrock")

Water relates

to the cosmos.

Water takes up a rhythm relating to the moon's cycle.

The tides of the seas are drawn higher on the shore

                                            and pull back more at the time

                                                           of the full moon.

Water is constantly recycled.

Streams and rivers flow to the seas.

Air gathers moisture and makes clouds that

produce rain,

returning the water to the land.

Water never really disappears.

It just keeps going round and round.

Water makes

the rainbows

that are seen

most often

in the

afternoon.

When sunlight is bent by raindrops

that act like prisms,

light breaks into colors.

What we see on earth,

the colors of the arch—red, orange, yellow, green, blue, and violet—

are reflections of the rainbow

which is actually round.

We see only a piece of the circle,

but the higher you are,

the more of the rainbow you will see.

Water seems to have

little power

of its own.

But it can wear down a mountain or a solid rock wall.

Water always finds crevices or the lowest point to work

or seep its way through.

Think of the Grand Canyon carved by the Colorado River,

deep in a gorge that runs from Utah to Arizona.

Water makes electric energy.

Held back by dams, water seeks

a way to fall to a lower level, generating power

through turbines. Most of the electricity

in our cities in the United States

comes from the energy generated by either water or coal.

Water has captured

the imagination

of poets,

philosophers,

and great teachers.

Lao-Tzu, the Chinese philosopher who founded Taoism around 600 B.C., used poetic language to speak of the power of water:

"Nothing on earth is so weak

and yielding as water,

Yet for breaking down that which is

firm and strong, it has no equal."

(*Tao Te Ching*)

Water plays a part

in many

religious

ceremonies.

Early Greek mythology gave personal names to water in a number of its forms. Oceanus was god of the ocean that surrounded the world. Poseidon (or Neptune) was the brother of Zeus and was given dominion over the sea. Aphrodite, goddess of beauty and all earthly joys, rose out of the sea on a half shell and was widely associated with the pleasure of bathing. You may have seen the glorious painting by Botticelli (the Italian painter of the 15th century Florentine School) of "The Birth of Venus," which is her Roman name. Even small rivers and streams were considered properties of various river nymphs. Thetis was a Nereid, or divine mermaid, and mother of Achilles, the great hero of the Trojan War. In Christianity, St. John the Baptist started the practice of baptism in water as a cleansing of sins. In some religious ceremonies, water is sprinkled when one is first given a name or, in the Muslim faith, when one marries.

Water is

both memory

and forgetting.

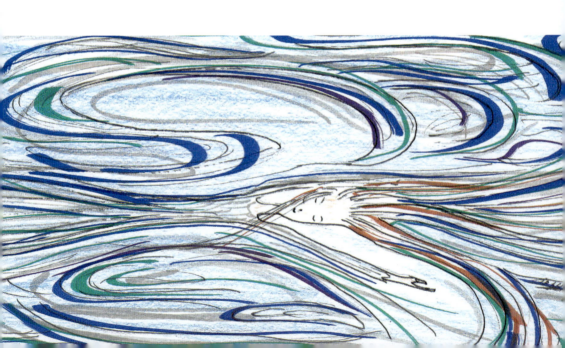

Heraclitus, in c. 500 B.C. wrote:

"It is not possible to step twice into the same river . . .

by the swiftness and speed of its change,

it scatters and collects itself again."

Water can purify itself.

Within only a few running feet,

a stream can become clear.

Lao-Tzu once more says:

"Who is there that can make muddy water clear?

But if permitted to remain still,

it will gradually become clear of itself."

(*The Way of Virtue*)

Gaston Bachelard, a philosopher in France, began to talk around 1940 about the elements, of which water is one, in a very special way. The other elements—earth, air, and fire—also received his attention, although he said that water was his favorite of the four. Of course, he wasn't the first to divide the essential components of the physical world in this way. Plato, and later in the Middle Ages, and later still in the Renaissance, philosophers had made this four-fold division. Bachelard is interested in the elements, not in their physical properties but in how they entice the imagination.

Bachelard asks his readers to look at each of them as real parts of the physical world, such as an actual pond or river, for example, but also as inner images in our mind's eye in daydreams or stories, like imagining the sea in *Moby Dick* or the "idle ship upon a painted ocean" in *The Rime of the Ancient Mariner*. Bachelard likes the way the imagination gathers many images around a central and essential one like water. He never saw the ocean until he was thirty years old, so he says that most of his examples are of streams, rivers, and lakes rather than the sea. In one of his observations he notes:

"The song of the river is . . .

cool and clear."

Bachelard prefers a life of joy and playfulness, saying that we are not meant to be full of unhappiness:

"Anguish is artificial:

we were meant to breathe freely."

But he reminds us that, on occasion, vital images such as water can also carry feelings of dread and fear. We can have too much of it or too little of it. Water in the oceans and even in lakes can be suddenly changeable when a storm blows up. A violent, stirred-up sea in a hurricane is terrifyingly powerful. On the other hand, with salt water all around a raft adrift on the sea, one can be horribly thirsty. The lack of clean water to drink is a big issue in our time in many parts of the world. We turn on the faucet and hardly know where the water comes from; many others have to carry it from distant locations.

Bachelard urges us

to take the lessons

of water to heart,

to look at things

in a watery way.

Deep water teaches us to think deeply,

to look beyond the surface of things.

Reflecting water teaches us to reflect first,

not always just to react quickly.

Reflecting is good

but thinking about or seeing only oneself

reflected in a pool, as did Narcissus in the myth,

is quite a deadly trap.

Actually, both birth and death relate to water.

Bachelard says that death associated with water

is more dream-like than death associated with earth,

and he reminds us of Ophelia's death among the flowers

in Shakespeare's *Hamlet*.

Life is often thought of as a sea journey.

Bachelard

calls

water

a "feminine"

element.

Why do you think this might be true?

Which other ones do you think are feminine?

Water is receiving and adaptable to other elements—

it is stirred by the wind or polluted by earth.

It has qualities like a mother giving birth,

not to a baby but to a spring that flows out of the ground.

In Bachelard's own words:

"We shall also see how profoundly *maternal* the waters are.

Water swells seeds and causes springs to gush forth.

Water is a substance that we see everywhere springing up

and increasing. The spring is an irresistible birth,

a *continuous* birth."

Water is an

in-between element,

in between fire

that puts it out

and earth that

needs it to sprout.

Bachelard talks about how fluid,

even flimsy, water images are:

"Images whose basis or matter is water

do not have the same durability as those of earth....

They do not have the vigorous life of fire images."

Water is a challenge.

It is not a natural medium for human beings, like air,

but one in which effort is required if we are to survive in it.

When we are swimming in the ocean,

we may experience a panicky moment when a large wave rolls in,

for example, but are thrilled when we dive under a wave

and come out on the top of the other side without getting rolled.

We may even feel momentarily that we have conquered

the sea and our fear of it.

Bachelard thinks that meeting challenges in the world

increases our feelings of power:

"More than anyone else, the swimmer can say:

the world is my will; the world is my provocation.

It is I who stir up the sea."

Water in rippling waves

that break on the shore

or water falling

from a waterfall

into a still pool—

in other words,

water that comes and goes—

imitates the

rhythm of the heart.

In these fluctuating motions,

water is like the beat of the heart or the rhythm of a poem.

        Bachelard says:

"Water, with its trickling and rushing,

with its ebb and flow,

        gives us the basic rhythms of life

and of language and poetry."

Watery sounds

flow through

the words that

we speak or

a paragraph

that is well written.

Bachelard speaks of this effect:

"Human language has a liquid quality,

a flow in its overall effect,

water in its consonants."

Bachelard ends his book on water

by talking about the thrill of hearing rushing water

flowing over stones in a stream.

He reminds us of how it brings joy even to a person

who is sad. With the words "trickling and rushing" or

"babbling and bubbling," the brook speaks its happy,

watery language to us.

The final word is Bachelard's:

"Not a moment will pass without repeating

some lovely, round word

that rolls over the stones."

Water is the beginning and the end.

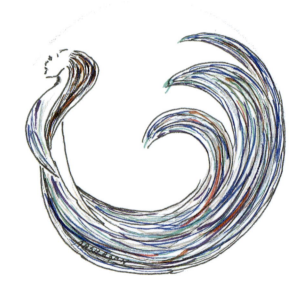

Water is the alpha and the omega

(the first and last letters of the Greek alphabet),

what unites heaven with earth

and with every living thing.

How vital it is!

About the author

Joanne H. Stroud, Ph.D., taught both literature and psychology at the University of Dallas before becoming a Founding Fellow of the Dallas Institute of Humanities and Culture. Currently on the faculty there, she teaches in the Spiritual and Cultural Psychology program. She is also Director of Publications and Editor of the Gaston Bachelard Translation Series, which consists of seven works on elemental imagination. She lectures in Dallas, New York City, and Connecticut and writes frequently on Bachelard, a 20th century French Philosopher of Science. In 2002 in Dallas she chaired a symposium on Bachelard, "Matter, Dream, and Thought," that attracted international attention. Her book, *The Bonding of Will and Desire*, was published in 1994.

Tashasan practices illustration and obscurity in the south of England.

Water Is Wet

Copyright © 2004 Joanne H. Stroud
Illustrations Copyright © 2004 Tashasan

ISBN 0-911005-46-3

All rights reserved. No part of this publication may be reproduced, stored in any retrieval system, or transmitted in any form or by any means, mechanical, photocopying, recording or otherwise, without permission in writing from the author, except by a reviewer, who may quote brief passages in a review to be printed in a magazine or newspaper.

The Dallas Institute Publications publishes works concerned with the imaginative, mythic, and symbolic sources of culture.

Publication efforts are centered at:
The Dallas Institute of Humanities and Culture
2719 Routh Street, Dallas, Texas 75201
www.dallasinstitute.org

Printed and bound in China
with Palace Press International

Design by Suzanna L. Brown
Illustrations by Tashasan

Second Printing, 2005